SPECIAL THANKS TO Melissa Darling, Chrysann Johnston, John Nicholson, Sabine and Felix, Chloe Rebisz, S. Nigel Rogers, Angela J. "Angie" Warren, Curtis Family, Avalon & Ansel, Griffin, Gwyneth, Garrett, and Gala Booher, Na'amah & Ellah Rosenzweig, Louis Garza, Doug Brownell, Ken McKnight, Gwen Daumit, The Olivotto Family, Aleria and Daelyn Stock, Luna Rose Eskew & Linley Sophia Eskew, Abhilash Lal Sarhadi, Voden Leger, Gryphon Leger, Larry Rosenthal, Harper Ray & Micah, The Andrews Family, The Smith Family, Edmund Patterson and Oliver Ingoldsby, Steve and Pam Solomon, Inara, Jaxon, Marge Rakow, Margaret Lowe, Rick Goheen, Aunt Diane, Rona & Menashe Dickman, Lenora Hans, Anthony, Eden & Harley, The Braskat-Arellanes Family, Catherine Durham, Rory Moore, Gaius Aileo, Alaric Aileo, Remington Aileo, Taylor Pedersen Smart, Faeryn Elara Duronslet, Eliyahu and Aharon, Seanna Lyon, Oscar Lyon, Phèdre, The Chatfields (Dan, Heather, & Ellie), Emma Catherine New, Ayres Family, Jahan Desai Siegal + Tilda Desai Siegal, Kimberly Kalani, Curtis Family, Amalia Rego, Robin Mayenfels, Aliyah Rose Siska, Huckleberry Jack Jacobs-Calicchia, Kindred Kids & Mogk Kids, Darwin James, Scarlett Danger, Abigail Karchin, Oliver Karchin, Oliver Belski, Dilara Kutay, Ela Kutay, Thomas and Draco Krautkramer, Declan & Wesley Carpenter, Joshua Erdin, Charlie Lovenstein, Simon Lovenstein, and Clark Mannion, John Boudreaux, Brielle Patterson, Greyson, Alayna, and Kyson Wilson, Aliyah Siska, and Jacob Steffen

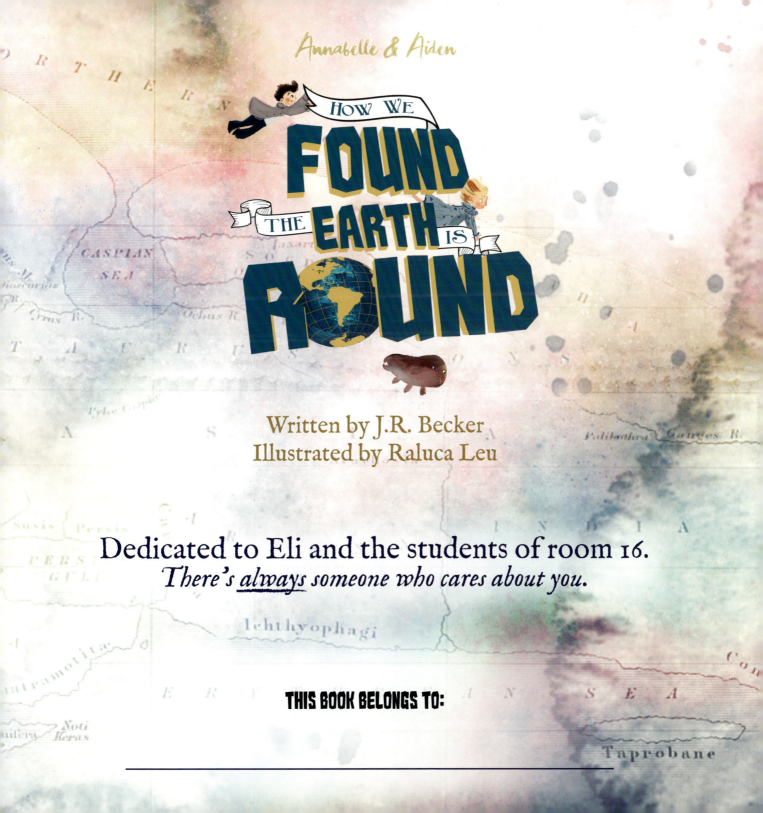

Annabelle & Aiden

HOW WE FOUND THE EARTH IS ROUND

Written by J.R. Becker
Illustrated by Raluca Leu

Dedicated to Eli and the students of room 16.
There's <u>always</u> someone who cares about you.

THIS BOOK BELONGS TO:

_____

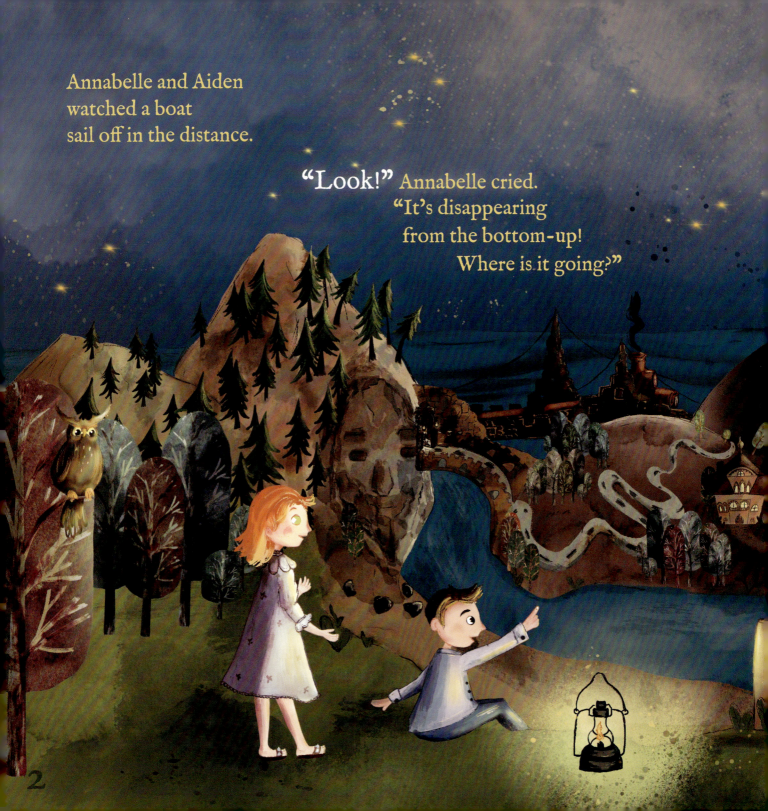

Annabelle and Aiden
watched a boat
sail off in the distance.

"Look!" Annabelle cried.
"It's disappearing
from the bottom-up!
Where is it going?"

2

"It's falling over the Earth's edge!" Aiden exclaimed.

"Falling?" Annabelle squinted to see. "Nah, it's going down too slowly for that."

Aiden thought some more.
"Maybe it's on... an elevator?"

"I've heard wilder things,"
said a voice.

4

"Tardigrade Tom!" the kids exclaimed,
"It's been a while.
Did you fall asleep again?"

"Oh, I just had a
quick two-year nap,"
Tom stretched out
his six stubby legs.

"Typical tardigrade,"
Aiden sighed.
"Well, we have to help that boat.
It's disappearing...over Earth's edge?
Does Earth even _have_ an edge?"

"Hop on my back," said Tom.
"To get answers, perhaps
we need to go back in time, to.... . . . .

Tardigrades (or "water bears") can pause their biological clocks.
They shrivel into a ball, dehydrate themselves, and lower their
metabolisms. They can exist like this for decades.

BABYLON
8,000 YEARS AGO

Annabelle and Aiden watched old men using clay to make an early world map.

It showed Earth as a flat disc surrounded by ocean.

The oldest world map ever found was made in Babylon in the 6th century BC — copied from a tablet made even 200 years earlier. It shows Earth as a disc surrounded by ocean and lands of magic and monsters.

"Beyond that," one of the men proclaimed, "lie the seven mysterious lands of monsters.

One home to a horned bull that attacks anyone nearby. Another as dark as the blackest black. A third with light brighter than any star... "

"This doesn't sound right," Tom said. He whizzed the kids forward in time.

7

In **EGYPT**, Pharaohs spoke of a flat Earth with Egypt at the center...

...all covered by a metal dome roof that was dotted with holes for stars, which were **lit up** by gods holding lamps from above.

With a flash, Tom zoomed them away again.

Annabelle and Aiden landed in ASIA —smack in the middle of an age-old argument.

The ancient Hindus said, "Earth is a *flat disc* on the back of four elephants standing on a turtle!"

Native American creation stories similarly say the Earth is created as soil piled on the back of a great sea turtle that grows until it is carrying the entire world.

The ancient Chinese replied, "Then how do you explain earthquakes? Earth must be a bowl on the back of a toad, and it shakes when the toad moves!"

It started getting heated, so Tom said, "Hold on, we're getting out of here."

In **GREECE**, early thinkers spoke of Earth as a flat disc with their city Athens in the center, surrounded by ocean, then other lands, and even more endless ocean.

"There are monsters there!" an explorer exclaimed, pointing at the distant lands. "I saw them! Monsters with dog's heads. And there's more: tigers with human faces! People with gigantic feet they use...as umbrellas!"

The Greek explorer Ctesias gave reports of encountering such monsters.

No matter where Annabelle and Aiden
went, people had very different beliefs
about the Earth's shape.

"But one thing is always the same,"
Annabelle noticed.

"They view themselves as the center of the world. The best. They're afraid of outsiders and see them as monsters."

"That sounds all too familiar," said Tom.

"Wait," said Annabelle. "I noticed something else. Take us back to Egypt, *but later in time*."

15

"Over there!"

Annabelle pointed to the Great Library of Alexandria.

The head librarian declared,

(AIR-UH-TOSS-THIN-EEZ)

"I, Eratosthenes, will map out the Earth's shape!"

Aiden laughed, "Impossible! Without computers?"

"Shh!" Annabelle said.
"He's going somewhere."

WHO GOT IT RIGHT FIRST?

Pythagoras is first credited with suspecting a spherical Earth in 500 BC. **Plato** learned it from him. **Aristotle** also wrote of a round Earth in *"On the Heavens"* in 350 BC.

17

Eratosthenes traveled to the nearby city of Syene. He came to a well and was surprised to see that the sunlight reached the bottom.

"There's no shadow," he thought, "so the sun must be directly above the well."

He noted it was twelve o'clock.

"Is the sun directly overhead at twelve o'clock in Alexandria too?" he wondered.

The next day he returned to Alexandria to see.

He didn't have a well, so he used a stick.

If the sun was straight above him, there would be no shadow.

SYENE

ALEXANDRIA

He waited until twelve o'clock. The sun *did* cast a shadow!

19

"So the sun is NOT directly overhead, like it is in Syene," he said.

"So the Earth can't be flat. It must be curved."

But he didn't stop there. He measured the shadow's angle at 7.2 degrees.

"I can use this measurement to discover the Earth's size: I just need to measure the distance from Syene to Alexandria."

The Nile flooded regularly, changing the landscape

Aiden laughed. "Preposterous!
They don't make rulers that long."

Annabelle said, "Look. He's hiring
a professional... step counter?"

They watched in amazement
as the counter walked and
walked and walked, counting
every   step.

              single

Long distances were measured by professional
distance walkers, who walked at a very regular
pace and counted each step.

around it, causing land ownership disputes. So these lands had to be re-surveyed annually after the flood.

One... two... three... four thousand... five thousand...

Tom explained, "He measured eight hundred kilometers. So Eratosthenes multiplied that distance *by fifty*."

"How'd he know it was *a fiftieth* of the entire Earth?" Aiden asked.

Tom said, "Because his stick cast a shadow of 7.2 *degrees*, which is *one-fiftieth* of a full circle, or 360 *degrees*."

"All this thinking is making me hungry," said Aiden. "Go on."

"Then imagine a pizza with fifty slices. If you knew the outer distance of one slice, you'd multiply that by fifty to find the size of the entire pizza. The distance from the well to the stick were like one slice. So he multiplied it by fifty to get ... forty thousand kilometers."

"The size of the whole Earth!"
the children yelled.

Tom smiled as he whizzed them back home.

The step counter's count of 800 kilometers was close. Today we know the distance to be 841 kilometers

7.2°

Eratosthenes's count of 40,000 km was close too. It's actually 40,075 km. He assumed that Earth is a perfect circle, but it actually has a bulge in the middle from its spin.

Aiden settled in the grass.

"So Eratosthenes figured out the shape and size of the entire Earth with nothing but a stick and some determination??"

"Pretty much," replied Tom. "Another Greek man named Aristotle figured it out too, just by looking at the Earth's shadow on the moon. Look at it now. See how the shadow is round?"

Aristotle had other reasons to suspect Earth was round as well. He saw stars in the sky in Egypt that he could not see when he was in distant lands. Also, he knew that gravity pulls equally in every direction from the center, which would make planets form as spheres. He assumed that Earth is a perfect circle, but it actually has a bulge in the middle from its spin.

The three friends watched the boat
disappear over the Earth's curve.

"There it goes," Annabelle smiled.
"With no need for a rescue. So what should we do now?"

"Y'know," Tom winked.

"If we fly up high enough, we might be
    able to see the Earth's curve ourselves."

27

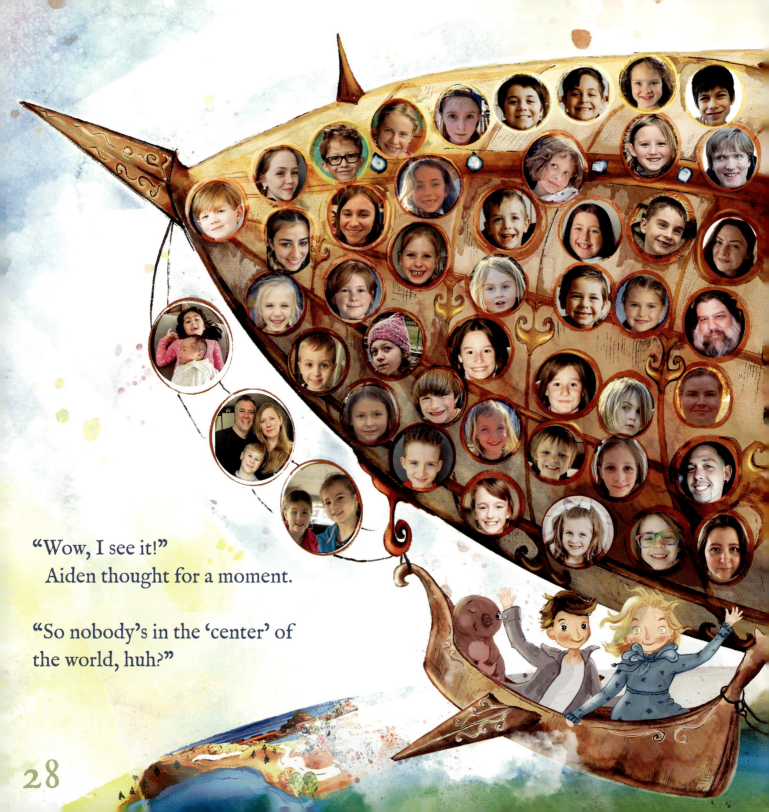

"Wow, I see it!"
   Aiden thought for a moment.

"So nobody's in the 'center' of
the world, huh?"

28

"That, or everybody is," Tom said.

Annabelle said, "Are there people who *still* think the Earth is flat?"

Tardigrade Tom winked. "Funny enough, they live *all around the globe*."

29

Grab Our Other Titles at *AnnabelleAndAiden.com*

"A beautiful, whimsical, and **deeply important book** for kids of all ages!"
- *Cara Santa Maria*, host of *Talk Nerdy* Podcast, co-author of *The Skeptic's Guide To The Universe*

"A great book. **Very smart. And kind.**"
— *Penn Jillette*, author of *God, No!*, comedian, magician of *Penn & Teller*

"Does an exemplary job of **explaining the origins of life** as we know it **to very young minds.** Heck, it could probably even teach a thing or two to their folks." - *Bill Nye Film*

"What a stunning & **refreshing set of books!** The exact words & tone to reach into a child's heart and brain and bring out such bliss & beauty. A wonderful addition to our family library!"
— *Mayim Bialik*, PhD, actress, author, and neuroscientist.

"**Beautifully illustrated** books for children that opens their minds and hearts to the wonders of science. Any child who reads it will find themselves **mesmerized, enlightened, and smiling.**"
- *Chip Walter*, National Geographic Fellow and author of *Last Ape Standing: The Seven-Million-Year Story of How and Why We Survived*